COUNTDOWN TO SPACE

APOLLO MOONWALKS

The Amazing Lunar Missions

Gregory L. Vogt

Series Advisor:
John E. McLeaish
Chief, Public Information Office, retired,
NASA Johnson Space Center

Enslow Publishers, Inc.
40 Industrial Road
Box 398
Berkeley Heights, NJ 07922
USA

PO Box 38
Aldershot
Hants GU12 6BP
UK

http://www.enslow.com

Library of Congress Cataloging-in-Publication Data

Vogt, Gregory.
Apollo moonwalks : the amazing lunar missions / Gregory L. Vogt.
p. cm. — (Countdown to space)
Includes bibliographical references and index.
Summary: Discusses the six Apollo missions that landed on the moon, describing the work performed there, what it is like to walk on the moon, and the collection of moon rocks.
ISBN 0-7660-1306-5
1. Projects Apollo (U.S.)—History Juvenile literature. 2. Moon—Exploration Juvenile literature. 3. Space flight to the moon Juvenile literature. [1. Project Apollo (U.S.)—History. 2. Moon—Exploration. 3. Space flight to the moon.] I. Title. II. Series.
TL789.8.U6A66323 2000
629.45'4'0973—dc21 99-16921
CIP

Printed in the United States of America

10 9 8 7 6 5 4 3 2 1

To Our Readers: All Internet addresses in this book were active and appropriate when we went to press. Any comments or suggestions can be sent by e-mail to Comments@enslow.com or to the address on the back cover.

Photo Credits: National Aeronautics and Space Administration

Cover Illustration: NASA (foreground); Raghvendra Sahai and John Trauger (JPL), the WFPC2 science team, NASA, and AURA/STScI (background).

Cover Photo: Astronaut James Irwin of Apollo 15 *on the Moon.*

CONTENTS

On the morning of July 16, 1969, the crew of Apollo 11 *blasted off launchpad 39A. Onboard were the two men who would soon become the world's first Moonwalkers.*

1

One Small Step

Neil Armstrong hesitated on the last rung of the ladder. He was wearing a bulky white space suit, and its gold-plated sun visor prevented anyone from seeing his face. Less than nine years after President John F. Kennedy announced that American astronauts would be landing on the Moon, Armstrong was about to make the historic first step onto the dusty surface of the Moon.

The television picture of the event that beamed 240,000 miles back to Earth was black and white. The picture looked fuzzy to the 600 million viewers on Earth, but the Moonwalker could clearly be seen.

Armstrong and his *Apollo 11* partner Edwin Aldrin, Jr., landed on the Moon on July 20, 1969. Their voyage began with the liftoff of their Saturn V rocket from the

Kennedy Space Center in Florida. The rocket, 363 feet tall, carried Armstrong, Aldrin, and Michael Collins to Earth orbit. Then, its third stage sent them speeding toward the Moon at 25,000 miles per hour. On the three-day trip to lunar orbit, the Moonwalkers prepared the landing craft (lunar module, or LM) they called the *Eagle*. It was a four-legged boxy-looking vehicle. It had a large rocket engine extending out of the bottom where its folded legs were attached. The bottom of the LM was the descent module. Connected to the top was a smaller box in which the crew would ride. This was the ascent module and it had its own rocket engine for returning from the Moon's surface.

After arriving in lunar orbit, Armstrong and Aldrin climbed into and sealed off the LM. By firing small direction control rockets, they slowly separated from the spacecraft where Collins remained. Collins was riding in *Columbia*, a cone-shaped capsule called the command module. It was attached to a cylindrical service module that had a large rocket engine at one end. These were the vehicles the astronauts would use to return to Earth.

Using the control rockets again, Armstrong aimed the LM so that its engine pointed in the direction they were moving. The LM's engine was then fired to slow the LM as it began its descent from orbit. The engine slowed the lander until its four legs touched the surface. Armstrong radioed back to Earth, "Houston, Tranquility Base here. The *Eagle* has landed."[1]

The lunar module, Eagle, *carried Neil Armstrong and Edwin Aldrin to the surface of the Moon.*

It was more than six hours after landing before the Moonwalkers were ready to explore the Moon. They had to put away all their food packages, flight plans, and other items they had needed. Then, they had to get into their space suits and remove the air from their LM.[2] Finally, Armstrong opened the hatch, stepped out, and began climbing down the ladder. He paused at the bottom rung and said, "I'm at the foot of the ladder. The LM footpads are only depressed in the surface about one or two inches, although the surface appears to be very, very fine grained as you get close to it. It's almost like powder." A few moments later Armstrong said, "I'm going to step off the LM now." He paused and then dropped the last few feet to the surface of the Moon. "That's one small step for a man, one giant leap for mankind."[3]

Fifteen minutes after Armstrong stepped onto the Moon, Aldrin followed him to the surface. The two Moonwalkers found that in spite of the heavy Earth weight of their space suits, their backpacks with an oxygen supply, and their own bodies (about three hundred sixty pounds total), they had no trouble moving about on the Moon. The Moon's gravity is only one sixth of Earth's gravity. Therefore, the astronauts with their suits and equipment weighed only about sixty pounds on the Moon. Armstrong and Aldrin discovered that the easiest

Astronaut Aldrin begins his climb out of the lunar module. Armstrong took this picture after he became the first man on the surface of the Moon.

way to get around was by bouncing from foot to foot with a kangaroo hop.

The main goal of the Moon landing was to collect samples of lunar rock and sediment to bring back to scientists on Earth. The astronauts also had experiments to set up. One task the Moonwalkers performed had nothing to do with science. They erected an American flag at the landing site. This turned out to be a difficult job.

The lunar sediment was only about six inches deep, and beneath it was solid rock. They hammered in the flagpole but it just would not stand up straight. Because there is not any atmosphere and therefore no wind on the Moon, a small telescoping rod was used to stretch the flag out. The rod would not work right, and the flag ended up with a permanent wave.[4] Michael Collins in *Columbia* did not have a television to view the Moonwalkers as viewers on Earth did. He radioed Mission Control in Houston for a description of what was happening. Mission Control responded, "They've got the flag up now and you can see the stars and stripes on the lunar surface." Collins replied, "Beautiful."[5]

The first walk on the Moon ended two and a half hours after it started. Aldrin reentered the lander first. Using ropes, he and Armstrong hoisted almost fifty pounds of lunar surface materials on board the ascent module. Armstrong then followed Aldrin inside. To make their vehicle lighter for the launch back to lunar orbit, the two Moonwalkers tossed their oxygen backpacks,

Armstrong and Aldrin proudly set up the American flag on the Moon.

Moon boots, used food packages, and other waste from the lander. Scientists on Earth reported that their seismometer, a device for measuring vibrations, was working better than they had expected. It measured the vibration of the Moon's surface when the debris from the lander hit it.[6] Then, they closed the hatch, stored their equipment, and rested. Later, after spending twenty-one hours on the Moon, the Moonwalkers used the descent module of the LM as a launch platform. A large rocket engine in the ascent module sent them back to join Collins in lunar orbit.

The *Apollo 11* landing was the start of a series of Apollo lunar expeditions. By the time the program ended three years later in 1972, twelve astronauts, all of them men, had walked on the surface of the Moon.

2

Preparing for the Moon's Surface

When the National Aeronautics and Space Administration (NASA) was given the challenge of sending astronauts to the Moon in less than a decade, the job seemed impossible. It was tough enough sending astronauts into Earth orbit, much less sending them all the way to the Moon, 240,000 miles away. If they ever got there, the astronauts would have to be protected from the dangers of the Moon. They would need some sort of protective clothing to keep them alive in a deadly environment.

Scientists already knew that there would not be any air on the Moon, and temperatures would range from 266°F in direct sunlight to –274°F in the shade.[1] There would be tiny high-speed micrometeorites striking the astronauts like little bullets. There was another problem as well. Some scientists believed the Moon's surface might be covered with thick layers of dust that could swallow up a Moonwalker like quicksand. The problems were tremendous. Eventually, more than three hundred thousand people teamed together to come up with solutions.

The problem of reaching the Moon was eventually solved by construction of the Saturn V rocket. Being able to survive while exploring the Moon's surface was accomplished with the construction of lunar space suits.

Building a space suit was a difficult job. It had to serve as a home away from home for the astronaut. It needed to contain air and exert pressure on the astronaut's body. Without air, the astronaut would suffocate. Without pressure, the astronaut's body would swell up and blood vessels inside the brain would break, causing a rapid death. Then, there was the problem of keeping the astronaut's body temperature at a comfortable level. Finally, the suit had to be flexible so the astronaut could walk around on the Moon. The suit had to have all its supplies, including oxygen and electricity, in some sort of backpack.

The suit that was finally constructed was an engineering marvel. It was a garment made up of twenty-one layers.[2] Each layer served a different function. Several layers were used as a pressure shell. They began with a rubber-coated fabric. Although it

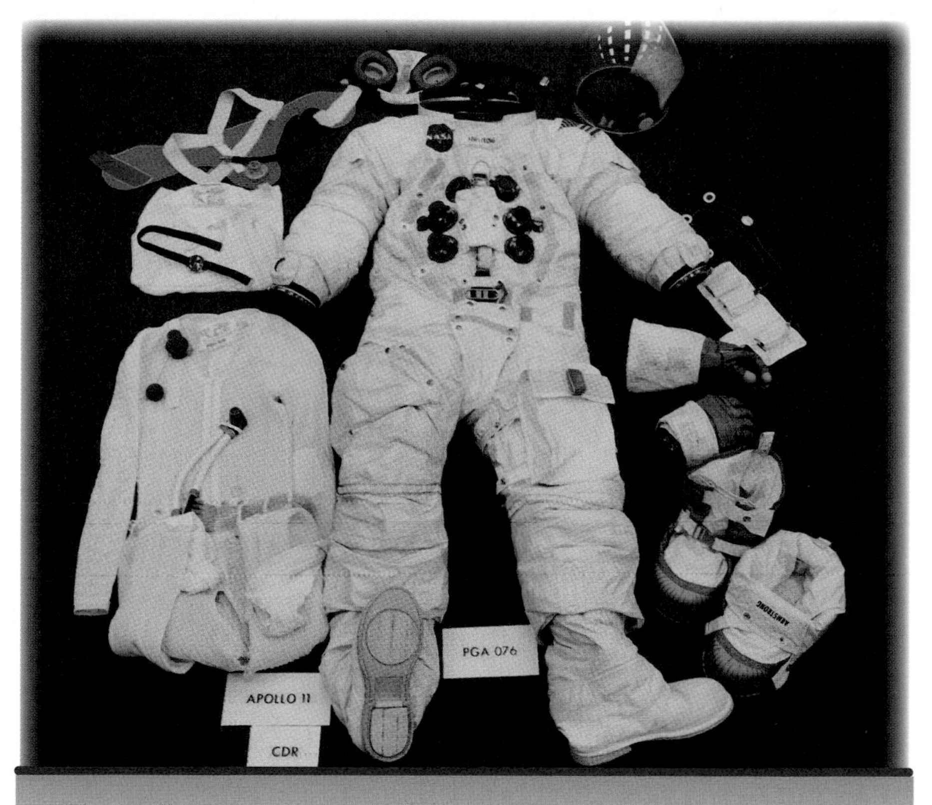

The Apollo space suit protected the astronauts from the conditions on the Moon. Without it, astronauts would not have survived.

could bend, the pressure shell could not stretch. Air pumped inside the shell was used for breathing and for pressure to keep the astronaut's body from swelling. These layers were covered on the outside by a tough fabric to prevent abrasion. On the inside, a soft fabric made the astronaut more comfortable. On top of the pressure garment were layers of insulation and finally a shell of bulletproof fabric to protect against the micrometeorites.[3]

Overshoes, like galoshes, were worn over the space suit's shoes to protect them from the hot rocks the astronaut would be walking on. Gloves sealed off the arms of the space suit, and a helmet with a gold-plated sun visor offered good visibility while protecting the astronaut's head. Oxygen, batteries, a cooling system, and a radio were contained in a backpack. Hoses from the pack carried oxygen and cooling water to connections on the front of the suit. Pockets in the sleeves and legs carried small tools. A plastic pouch in one leg with a tube connected to the astronaut was used in case he had to go to the bathroom.

The twenty-one layers of the Moon suit were not the only things the astronauts wore on the Moon. Before putting on the space suit, they put on underwear that resembled long johns, and a garment consisting of stretchy fabric laced with tubes. The tubes were connected to a water supply that would circulate around the astronaut to keep him cool.[4]

3

Training for the Moon

Since no one had ever been to the Moon, getting astronauts ready for walking on its surface was a matter of guesswork. First, the astronauts had to be selected. What kind of persons should they be? Most people thought that the best candidates for going to the Moon would be test pilots. After all, the first task was to fly to the Moon and land safely. Test pilots were our country's most experienced pilots. The candidates would have to be shorter than five feet, eleven inches in order to fit comfortably inside spacecraft.[1]

Astronaut candidates had to pass many fitness tests. How long could they run on a treadmill? How long could they keep their feet in ice water? How many balloons could they blow up before collapsing? Eventually, several groups of astronauts were selected for the Apollo missions to the Moon.

Training took on many forms, the most important of which was flying in space. Individual astronauts first flew into space in Mercury space capsules. The capsules orbited Earth a few times and then splashed down in the ocean. The Mercury missions demonstrated that humans could survive and work in space.

The Gemini missions followed Mercury. Two astronauts flew together in slightly larger space capsules. They experimented with spacewalking and docking two spacecraft in orbit. Both tasks would be essential in traveling to and from the Moon safely.

Finally, the early Apollo missions made sure that the Apollo rockets and spacecraft worked. The mission to walk on the Moon was moving rapidly—too rapidly, as it turned out. During a January 1967 training exercise inside an Apollo space capsule on top of a Saturn I rocket, astronauts Edward H. White II, Virgil I. Grissom, and Roger B. Chaffee were tragically killed in a flash fire. The accident investigation discovered that some basic safety procedures had been overlooked. This led to important changes in the safety features of the Apollo space capsule.[2] It would be two years before the program was back on track.

In a rapid series of missions, Apollo spacecraft and Saturn boosters were tested in space. Astronauts on *Apollo 7* and *Apollo 9* tested spacecraft in Earth orbit.

Apollo 8 and *Apollo 10* astronauts traveled to orbit the Moon and back. Each mission was more ambitious than the last. By July 1969, all systems were ready for an attempted Moon landing.

Meanwhile, astronauts had to learn how to function inside their space suits. With all the layers, moving about would be a challenge. NASA provided an aircraft, called the KC-135, that had a padded interior. With the aircraft flying in special arcs through the sky, astronauts inside could get used to the floating that would take place on the way to and from the Moon. Wearing space suits with extra weights attached to them, astronauts drifted underwater in a pool at the Johnson Space Center in Houston, Texas. The floating experience helped them develop techniques for using tools and moving about in space.[3]

Astronaut Alan L. Bean spends some time in a lunar surface training machine. The simulator imitates the one-sixth gravity conditions on the Moon.

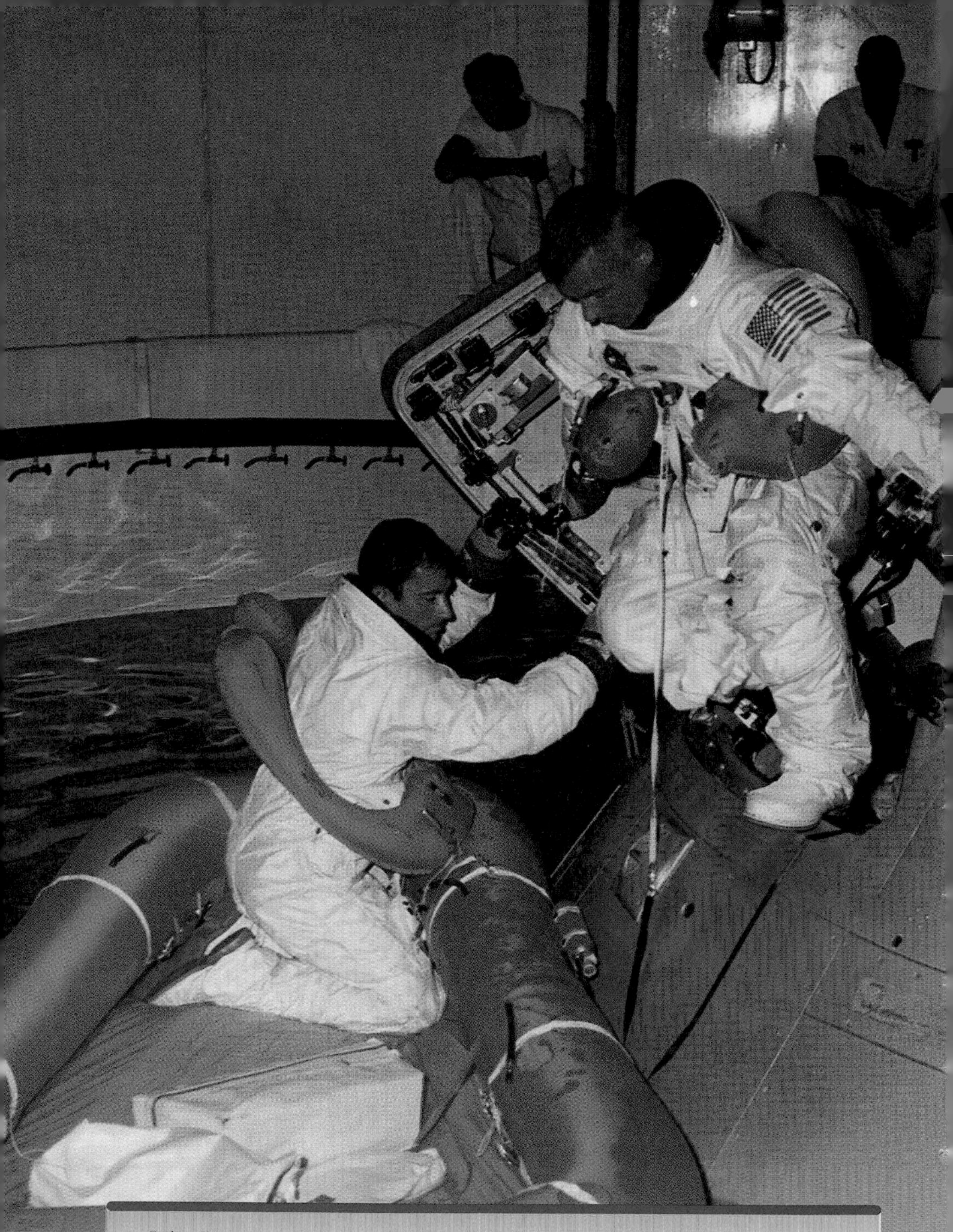

John Young and Eugene Cernan train in the pool at the Johnson Space Center. They are practicing getting out of the command module that will take them to and from the Moon.

Another important kind of training was in geology. On the Moon, the astronauts would have to photograph and select rock and sediment samples to bring home. Since they could bring back a limited number of samples, which ones would they choose? Geologists took the astronauts on geology field trips to the Grand Canyon where they practiced making observations. The astronauts also went to Taos, New Mexico, to observe the Rio Grande gorge, which would possibly be similar to a Moon valley called the Hadley Rille.[4] Since the Moon's surface is made of volcanic rock, astronauts traveled to the volcanic islands of Hawaii and to a volcanic region in Arizona. There, scientists exploded dynamite to create Moonlike craters across the surface of a dusty lava plain. Future Moonwalkers trudged over craters, plucked up interesting rocks and soil samples, and

Edwin Aldrin and Neil Armstrong study rocks during a geological field trip in Texas.

practiced operating their scientific and photographic equipment.

While the astronauts were training and the engineers were finishing the space vehicles and tools, another group was working on the science experiments to be conducted on the Moon. Going to the Moon was a great scientific opportunity. Scientists thought of experiments that the astronauts could do on the Moon. Automatic equipment was developed that could radio the experimental data to Earth after the astronauts had gone. One experiment was designed to collect particles ejected from the Sun. Another experiment involved a laser reflector aimed toward Earth. When a laser beam sent from Earth bounced off the reflector, the length of time the beam traveled could be used to determine the exact distance between the Moon and Earth. Scientists also designed a seismometer to be placed on the Moon to measure Moonquakes and impacts of meteorites striking the Moon.[5]

4

Rocks, Dust, and Craters

With the return of the *Apollo 11* crew from the Moon, scientists from around the world were delighted. Except for a few meteorites, the forty-six pounds of lunar rock and sediment the astronauts had collected were the first materials from outer space they could study. However, the *Apollo 11* material was from just one location on the Moon—the broad lava plain called Sea of Tranquility.[1] Scientists wanted material from other locations on the Moon, such as the lunar highlands. Other missions were needed.

Four months after the *Apollo 11* crew returned home, Alan Bean and Charles Conrad of the *Apollo 12* crew climbed down the ladder of the lander craft they named

Charles Conrad, Richard Gordon, and Alan Bean were selected for Apollo 12, *the second Moon landing mission. Bean and Conrad completed two Moonwalks in eight hours.*

Intrepid and stepped out onto the Ocean of Storms region of the Moon.

Armstrong's and Aldrin's expedition on the Moon lasted only two and a half hours. Only a small part of their time was spent conducting geologic surveys of the Moon. Bean and Conrad stayed long enough to conduct two Moonwalks, lasting a total of almost eight hours. Most of their time was spent collecting samples and studying the Moon's geology.

Geologists on Earth were delighted with the crew's running commentary on what the Moon looked like. They described the region as sort of a rolling plain and talked about rocks that were gray, tan, and brown. They spoke of rock-rimmed craters in all directions and boulders as wide as twenty feet.[2]

The first thing Bean and Conrad did on their initial Moonwalk was get used to walking on the Moon. Every movement they made stirred up fine lunar dust. In the low gravity, it was easy to kick up the dust. It fell back slowly, covering equipment the crew had to set up. They unloaded the lander with pulleys and ropes and complained they were getting dirty.

Once they adjusted to moving about, the astronauts opened a radio antenna and aimed it at Earth. They also set up several scientific experiments that measured the very thin lunar atmosphere and lunar quakes, and studied magnetic fields. One very important job the astronauts had was to remove a radioactive plutonium rod from a canister and place it into a power center. The heat from the radioactivity would be converted into electricity for running experiments. However, the plutonium rod was stuck.

"I tell you what worries me, Pete," said Bean. "If I pull on it too hard, it's a very delicate lock mechanism. . . . I just get the feeling that it's hot and swelled in there or something. Doesn't want to come out. I can sure feel the heat, though, on my hands. Come out of there! Rascal."[3]

Alan Bean sets up the Apollo Lunar Surface Experiments Package.

The crew had to strike the canister several times with a hammer to free the rod. With the rod in place, the power center began generating electricity and the experiments began working.[4]

As Bean and Conrad prowled around their landing site, they often stopped to pick up interesting rocks and sediment to bring home. "I can see everything from fine-grain basalt—as I come running across the area here—to a few fair(ly) coarse-grain ones," Bean said. "I don't know really what I would call it. . . . It looks almost like a granite, but of course it probably isn't, but it has the same sort of texture."[5] Some of the rocks had small crystals and other rocks had glass on the surface. The glass came from meteorites impacting the Moon's surface. The heat generated by the impacts caused those portions of the Moon's surface to melt. Some of the molten material splashed onto other rocks, forming a glassy coating. In addition to collecting samples, Bean and Conrad dug small trenches to see what the subsurface of the Moon looked like. They pounded in core tubes, like apple corers, to bring up samples from beneath the surface.[6]

After resting on the lander, the two Moonwalkers visited the robot *Surveyor 3* spacecraft that had landed on the Moon in 1967. The spacecraft rested on the southern rim of a large crater. Bean and Conrad noted the dust coating on the spacecraft and even a footprint made by the lander's feet where it had bounced during touchdown. "Hey, we got a nice brown *Surveyor* here,"

The unmanned Surveyor 3 *spacecraft landed on the Moon in 1967. Astronaut Charles Conrad examines the* Surveyor's *camera. He brought it back to Earth for examination. The lunar module is in the background of the photograph.*

said Bean, referring to its dust coating.[7] Using a cutting tool, the crew snipped off some of *Surveyor*'s tubing, its trenching scoop, and a camera. The souvenirs would be returned to Earth for scientists to see how well the material survived thirty-seven months on the Moon.[8]

Like the *Apollo 11* crew before, Alan Bean and Charles Conrad blasted off the Moon's surface in the upper stage of their landing craft. Their take of lunar materials was 112 pounds!

5

Moon Buggy

The *Apollo 13* mission was scheduled to travel to the Fra Mauro region of the Moon. The crew included James Lovell, John Swigert, and Fred Haise. It was a Moonwalk that was never to be. On the way to the Moon, an oxygen tank in the service module of their spacecraft exploded. The Moon landing was cancelled, and the crew had to make an emergency return to Earth.

Returning to the Moon was now the job of *Apollo 14*. Alan Shepard, the first American to travel into space, led the crew. Stuart Roosa and Edgar Mitchell joined him. Shepard and Mitchell touched down in the Fra Mauro region of the Moon.

During two Moonwalks, Shepard and Mitchell stayed out a total of nine hours twenty-two minutes.

They used a two-wheeled cart to help them carry tools, experiments, and samples long distances on the Moon. The astronauts liked to refer to the cart as the "rickshaw," "wheelbarrow," or "caddy cart."

Like previous Moon-landing crews, the *Apollo 14* crew had experiments and cameras to set up and antennas to point toward Earth. Again there was a seismometer to measure lunar quakes. A string of microphones was placed on the Moon's surface to hear vibrations when

Apollo 14*'s crew was Stuart Roosa, Alan Shepard, and Edgar Mitchell. They used the two-wheeled "rickshaw" (inset) to move samples and equipment during their Moonwalks.*

the crew fired the thumper. The thumper was a device that Mitchell pressed to the ground while triggering a small explosive charge. "This thing has a pretty good kick to it. . . . Sort of like firing both barrels of a 12-gauge shotgun, and at once," Mitchell said.[1] The vibrations from the explosions told geologists on Earth about the structure of the Moon's interior.

During their second Moonwalk, Shepard and Mitchell dragged the rickshaw over thirty-six hundred feet from their lander in an attempt to reach the thirteen hundred-foot-wide Cone crater. They got close but never actually reached the crater rim. To reach the rim, they had to climb a steep sandy slope. Their feet sunk and they kept sliding back. Because their oxygen supply was being used up too quickly, the attempt was cancelled by Mission Control on Earth. Nevertheless, the two astronauts examined many small craters as they made their way around automobile-sized boulders—some twice as tall as the Moonwalkers.[2]

Near the end of the second Moonwalk, Shepard surprised mission controllers on Earth. He had had some tool engineers at the Johnson Space Center create a golf club head to attach to one of his sampling tools. Shepard announced,

> You might recognize what I have in my hand as the handle for the contingency sample return; it just so happens to have a genuine six iron on the bottom of it. In my left hand, I have a little white pellet that's familiar to

> millions of Americans. I'll drop it down. Unfortunately, the suit is so stiff, I can't do this with two hands, but I'm going to try a little sand trap shot here.

He then took a one-handed swing at the ball. He topped the ball and buried it in the lunar dust. Edgar Mitchell proclaimed, "You got more dirt than ball that time." Eventually Shepard sent a second ball flying. As he watched it disappear, he bragged that it went for "miles and miles and miles."[3]

While the two-wheeled cart enabled the *Apollo 14* Moonwalkers to go farther from their lander than the *Apollo 11* or *Apollo 12* crews, travel was still exhausting for the crew. This was not true for the *Apollo 15* crew. They had a special treat—an electric car to ride in.

Apollo 15 *astronaut James Irwin gives a military salute while standing next to the American flag. To the right is the lunar roving vehicle that Irwin and Scott used to explore the Moon's surface.*

Just ten days after the second anniversary of the *Apollo 11* Moon landing, the lunar lander *Falcon* touched down in the Hadley Rille, Apennine Mountain region of the Moon. David Scott and James Irwin were on board. This crew made three Moonwalks lasting a total of eighteen hours thirty-five minutes.

After the usual set-up activities, Scott and Irwin unfolded an electric car—called the lunar roving vehicle—from the side of the lander. The car had four wire-mesh wheels, each powered by an electric motor. It had two seats and a place to carry up to 127 pounds of equipment and samples. Battery power on the rover permitted the crew to travel many miles from the landing site without tiring out. As they took the rover for a test drive, Scott described its handling: "There doesn't seem to be too much slip. I can maneuver pretty well with the thing. If I need to make a turn sharply, why, it responds quite well. There's no accumulation of dirt in the wire wheels."[4]

The landing site of *Falcon* was especially interesting. Within a short drive of the lander, there was a lava plain, a canyon, and the slopes of a mountain region. Hopping in the rover, Scott and Irwin drove to the rim of the Hadley Rille. This meandering canyon on the Moon was probably a channel through which molten lava once moved. It is about one mile wide and ranges from six hundred to twelve hundred feet deep. Scott described the scene: "I can see from up at the top of the rille down, there's debris all the way."[5]

During their second Moonwalk, Scott and Irwin drove to the slopes of the Apennine Mountains. The mountains, towering up to fifteen thousand feet above the surrounding lava plain, were too tall for the crew to climb, even with the rover. They were able to drive up to the base of the slopes to collect samples and take pictures. In the third Moonwalk, the crew explored about eight miles of the rim of Hadley's Rille.[6]

Scott and Irwin traveled across the Moon at speeds up to eight miles per hour, stopping frequently to take samples. Scientists were especially eager for the samples

The battery-powered rover helped astronauts Scott and Irwin travel far from their landing site without getting tired.

this crew would collect. All the previous samples were collected from lava plains and were dark and dense. Scientists thought the rock was from 3.5 to 4 billion years old. Since the lava had to flow across something, there had to be older rocks on the Moon. Scientists believed that samples taken from the mountains would be the older rocks and they would be lighter in color and less dense. The crew was instructed to look carefully for different kinds of rocks. David Scott found a light-colored rock and he radioed Earth: "I think we got what we came for."[7]

Collecting rocks was not all the crew members accomplished. They also hammered core tubes into the Moon as deep as 7.75 feet. Scientists later counted forty-four layers in the cores. That meant the cores represented dust that had been kicked up during forty-four separate volcanic eruptions and meteorite impacts in the lunar past.

Lunar roving vehicles were carried by the last two astronaut crews to visit the Moon. *Apollo 16*'s John Young and Charles Duke landed the *Orion* at the Descartes region of the Moon. Seven months later, *Apollo 17*'s Eugene Cernan and Harrison Schmitt landed *Challenger* in the Taurus-Littrow region. Young and Duke used their rover to travel seventeen miles during three Moonwalks, and Cernan and Schmitt traveled a record of twenty-two miles during their three Moonwalks.[8]

Like the previous lunar crews, *Apollo 16* crew members Young and Duke set up many experiments. By now there were twenty automatic experiments on the Moon, transmitting data to Earth.

By the end of the three *Apollo 16* Moonwalks, the two astronauts had spent a record of twenty hours fifteen minutes on the lunar surface. They collected 210 pounds of samples and took over 10,800 pictures.[9] In one of the highlights of their explorations, they drove their rover up a twenty-degree slope of Stone Mountain.

Charles Duke of Apollo 16 *collects lunar samples at the Descartes landing site.*

The crew named the mountain after a mountain in Georgia. They reached a terrace about seven hundred feet above their landing craft *Orion*. Duke exclaimed "Wow! What a place! What a view, isn't it, John?"

"It's absolutely unreal!" Young replied. Duke continued with his excitement. "It's just spectacular. Gosh, I have never seen . . . All I can say is 'spectacular,' and I know y'all are sick of that word, but my vocabulary is so limited."[10]

The thing that distinguished the *Apollo 17* landing from the rest was that one of its crew members, Harrison Schmitt, was not a test pilot before becoming an astronaut. Schmitt was a geologist. He brought a new perspective to the exploration—the eye of a scientist. Cernan's and Schmitt's explorations carried them farther and longer across the Moon's surface. They spent more time exploring than the crews of *Apollo 11*, *12*, and *14* combined.

They also ran into their share of difficulties on the Moon. Cernan accidentally knocked the rear fender off their lunar rover, and the wheel sprayed dust all over them and their equipment as they drove. During their second Moonwalk, following instructions from engineers at Houston, they made a new fender out of some maps and tape. It was the first automotive repair on the Moon.[11]

Schmitt's geology background enabled him to not only spot unusual rocks and sediment on the Moon but

The repaired fender on the lunar rover is the result of the first automotive repair on the Moon!

also to understand the significance of what he was seeing. After adjusting a camera on the rim of the crater Shorty, he spoke excitedly—"There is orange soil!"[12] Cernan came up and agreed with Schmitt. The orange soil turned out to be glassy particles produced by volcanic activity. Samples of the soil and other discoveries they made went a long way in enabling scientists to interpret the Moon's long history.

At exactly 12:40:56 A.M. eastern standard time on

Geologist-astronaut Harrison Schmitt stands next to a huge boulder on the Moon.

December 14, 1972, the last walk on the Moon ended.[13] Cernan turned for one last look at the Moon from the surface and radioed Earth: ". . . I'm on the surface . . . as we leave the Moon at Taurus-Littrow, we leave as we came, and, God willing, we shall return with peace and hope for all mankind."[14]

6

Returning to the Moon

When the *Eagle* landed on the Moon with the *Apollo 11* crew, it carried a small plaque attached to one of its landing legs. The plaque read, "HERE MEN FROM THE PLANET EARTH FIRST SET FOOT UPON THE MOON JULY 1969 A.D. WE CAME IN PEACE FOR ALL MANKIND."[1] Three and a half years later the last crew left a plaque on the Moon attached to the legs of the *Apollo 17 Challenger* landing craft. Their plaque read, "HERE MAN COMPLETED HIS FIRST EXPLORATION OF THE MOON DECEMBER 1972 A.D. MAY THE SPIRIT OF PEACE IN WHICH WE CAME BE REFLECTED IN THE LIVES OF ALL MANKIND."[2]

From 1969 to 1972, six lunar landers touched down on the Moon, twelve astronauts walked on its surface, and about eight hundred fifty pounds of lunar rock and

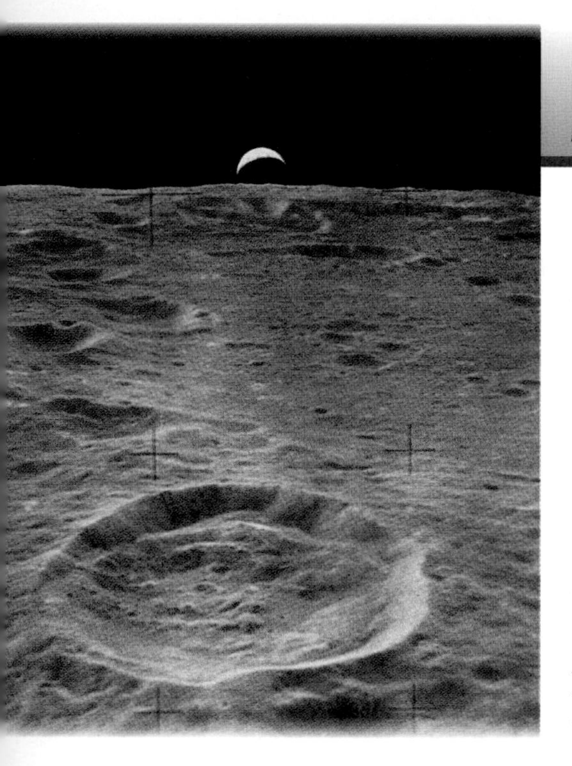

Samples from the Moon's surface helped scientists date the Moon. It is probably more than 4 billion years old.

sediment were brought home. What was learned by all that effort?

Analysis of the lunar samples told scientists that the Moon is very old and probably was formed when Earth was struck with a small planet over 4 billion years ago. The debris kicked up from the collision formed a disk that circled Earth and eventually collected in one clump to become the Moon. Scientists learned that the dark, flat areas of the Moon are made from dark volcanic rocks that cooled on the Moon's surface. The light, mountainous areas are made from light-colored rocks that cooled in the Moon's interior. None of the samples showed any signs of present life or fossils of past life. Instruments left on the Moon helped scientists monitor the Moon's temperature, heat flow from the interior, and Moon quakes. The data, radioed to Earth, showed that the Moon is divided inside in layers and it has no magnetic field. Magnetic fields are generated in planets by molten cores. The finding showed that the Moon is solid all the way through.

An important discovery about the Moon was what it is made of. Moon rocks contain iron, titanium, aluminum, silicon, and oxygen. These are materials that could be quite useful when we go back to the Moon someday. The materials can be refined to create metals and glass to build permanent bases on the Moon, as well as oxygen for breathing.[3]

More recent discoveries about the Moon make moving there even more possible. Satellite observations

Someday, oxygen may be mined on the Moon so that people can live there permanently.

of the Moon by the *Lunar Prospector* mission have shown that water ice may be found in the dark areas near both poles.[4] The ice could be melted for drinking water and for making rocket fuel.

Unfortunately, there are no firm plans to return to the Moon. The nations of the world that explore space are busy constructing a giant space station that will be completed some time early in the twenty-first century. They will use it to do scientific research and to learn about living in space for long periods. Eventually, the station will become a jumping off place for human exploration of our solar system. The first stop will probably be the Moon, and then it will be the planet Mars and beyond.

When future astronauts do return to the Moon, they will not come back as explorers. They will not stuff their spacecraft with rocks and sediment, and then turn around and come back in a few days. They will go to the Moon to stay. They will use their landing craft as temporary living quarters while they construct permanent scientific research stations on the Moon. Later, they will build cities. These new astronauts will become the first space pioneers.

Apollo Moon Landing Mission Summary

Mission	Launch Date	Crew	Lander Name	Location and Samples
Apollo 11	July 16, 1969	Neil A. Armstrong (L) Edward E. Aldrin (L) Michael Collins*	*Eagle*	Sea of Tranquility; 46 pounds
Apollo 12	Nov. 14, 1969	Charles Conrad, Jr. (L) Alan L. Bean (L) Richard F. Gordon, Jr.*	*Intrepid*	Ocean of Storms; 75 pounds
Apollo 14	Jan. 31, 1971	Alan B. Shepard, Jr. (L) Stuart A. Roosa (L) Edgar D. Mitchell*	*Antares*	Fra Mauro; 95 pounds
Apollo 15	July 26, 1971	David R. Scott (L) James B. Irwin (L) Alfred M. Worden, Jr.*	*Falcon*	Hadley-Apennine; 169 pounds
Apollo 16	April 16, 1972	John W. Young (L) Charles M. Duke, Jr. (L) Thomas K. Mattingly II*	*Orion*	Descartes; 210 pounds
Apollo 17	Dec. 7, 1972	Eugene A. Cernan (L) Harrison H. Schmitt (L) Ronald E. Evans*	*Challenger*	Taurus-Littrow; 258 pounds

L=Lander crew

*Command module pilot who did not walk on the Moon

CHAPTER NOTES

Chapter 1. One Small Step

1. *Apollo 11 Technical Air-to-Ground Voice Transcription* (Houston: Manned Spacecraft Center, July 1969).

2. Ibid., p. 6.

3. *Lunar Surface Journal—Apollo 11*, n.d., <http://www.hq.nasa.gov/office/pao/History/alsj/frame.html> (April 2, 1999).

4. Edgar M. Cortright, ed., *Apollo Expeditions to the Moon* (Washington, D.C.: Scientific and Technical Information Office, National Aeronautics and Space Administration, 1975), pp. 214, 216.

5. *Lunar Surface Journal—Apollo 11.*

6. Cortright, p. 219.

Chapter 2. Preparing for the Moon's Surface

1. Michael A. Seeds, *Foundations of Astronomy*, 3rd ed. (Belmont, Calif.: Wadsworth Publishing Company, 1992), p. 462.

2. Lillian D. Kozloski, *U.S. Space Gear, Outfitting The Astronaut* (Washington, D.C.: Smithsonian Institution Press, 1994), pp. 87–88.

3. Ibid., p. 88.

4. Ibid., pp. 84–85.

Chapter 3. Training for the Moon

1. Edgar M. Cortright, ed., *Apollo Expeditions to the Moon* (Washington, D.C.: Scientific and Technical Information Office, National Aeronautics and Space Administration, NASA-SP-350, 1975), p. 146.

2. Ibid., pp. 66, 73–75.

3. Ibid., p. 138.

4. Ibid., p. 139.

5. Ibid., pp. 238–241.

Chapter 4. Rocks, Dust, and Craters

1. Linda N. Ezell, *NASA Historical Data Book*, vol. 3 (Washington, D.C.: National Aeronautics and Space Administration, 1988), p. 85.

2. *Apollo 12, A New Vista for Lunar Science* (Washington, D.C.: National Aeronautics and Space Administration, EP-74, 1970), p. 5.

3. *Lunar Surface Journal—Apollo 12*, n.d., <http://www.hq.nasa.gov/office/pao/History/alsj/frame.html> (April 2, 1999).

4. *Apollo 12, A New Vista for Lunar Science*, p. 8.

5. *Lunar Surface Journal—Apollo 12*.

6. *Apollo 12, A New Vista for Lunar Science*, p. 11.

7. *Lunar Surface Journal—Apollo 12*.

8. *Apollo 12, A New Vista for Lunar Science*, p. 13.

Chapter 5. Moon Buggy

1. *Lunar Surface Journal—Apollo 14*, n.d., <http://www.hq.nasa.gov/office/pao/History/alsj/frame.html> (April 2, 1999).

2. Walter Froehlich, *Apollo 14: Science at Fra Mauro* (Washington, D.C.: National Aeronautics and Space Administration, NASA-EP-91, 1971), pp. 6–9.

3. *Lunar Surface Journal—Apollo 14*.

4. *Lunar Surface Journal—Apollo 15*, n.d., <http://www.hq.nasa.gov/office/pao/History/alsj/frame.html> (April 2, 1999).

5. Ibid.

6. *Apollo 15, At Hadley Base* (Washington, D.C.: National Aeronautics and Space Administration, NASA-EP-94, 1971), pp. 2, 31.

7. Ibid., p. 3.

8. Linda N. Ezell, *NASA Historical Data Book*, vol. 3 (Washington, D.C.: National Aeronautics and Space Administration, 1988), p. 89.

9. Ibid.

10. *Lunar Surface Journal—Apollo 16*.

11. David Anderton, *Apollo 17, At Taurus-Littrow* (Washington, D.C.: National Aeronautics and Space Administration, NASA-EP-102, 1973), pp. 9–10.

12. Ibid., p. 13.

13. Ezell, pp. 92.

14. Anderton, p. 19.

Chapter 6. Returning to the Moon

1. Russell E. Chappell, *Apollo* (Washington, D.C.: National Aeronautics and Space Administration, EP-100, 1974), p. 33.

2. David Anderton, *Apollo 17, At Taurus-Littrow* (Washington, D.C.: National Aeronautics and Space Administration, NASA-EP-102, 1973), p. 19.

3. *A 3-Minute Guide to the Moon*, Houston, Tex.: Lunar and Planetary Institute, <http://cass.jsc.nasa.gov/expmoon/3minmoon.html> (January 15, 1999).

4. Ibid.

GLOSSARY

Apollo—The NASA missions in which astronauts traveled to the Moon and back.

ascent module—The upper stage of the lunar lander that carried the astronauts to the Moon's surface and back into lunar orbit.

command module—The cone-shaped spacecraft that carried astronauts to and from Moon orbit.

descent module—The bottom stage of the lunar lander.

Gemini—A NASA space mission in which two astronauts orbited Earth to practice techniques that would be needed for going to the Moon.

lunar rover—Electric vehicle for riding on the Moon's surface.

Mercury—A NASA mission in which one astronaut orbited Earth to learn how to live and function in space.

National Aeronautics and Space Administration (NASA)—The agency of the United States government that explores outer space.

orbit—The path a spacecraft or satellite travels when it circles a planet or a moon.

Saturn V—The giant rocket used to send astronauts to and from the Moon.

sediment—Pebbles, sand, and dust worn down from rock.

seismometer—A scientific instrument for measuring the quaking of a planet when its rock moves against itself or when the planet is struck by an object such as a meteorite.

service module—The cylindrical stage of the Apollo spacecraft that provided electrical power and carried oxygen and a rocket engine for the return from the Moon.

Surveyor—A robot spacecraft that landed on the Moon to allow scientists to learn about its surface before astronauts traveled there.

FURTHER READING

Books

Cole, Michael D. *Apollo 11: First Moon Landing*. Springfield, N.J.: Enslow Publishers, Inc., 1995.

———. *Moon Base: First Colony in Space*. Springfield, N.J.: Enslow Publishers, Inc., 1999.

Donnelly, Judy. *Moonwalker: The First Trip to the Moon*. New York: Random House, 1989.

Kennedy, Gregory. *Apollo to the Moon*. Broomall, Pa.: Chelsea House Publishing, 1992.

Vogt, Gregory. *Apollo and the Moon Landing*. Brookfield, Conn.: The Millbrook Press, 1991.

Internet Addresses

Dunbar, Brian. *NASA Homepage*. "Search the NASA Web." September 13, 1999. <http://www.nasa.gov/> (September 13, 1999).

Johnson Space Center. "Earth from Space." August 26, 1999. <http://earth.jsc.nasa.gov/> (August 26, 1999).

National Aeronautics and Space Administration. *NASA Human Spaceflight*. September 10, 1999. <http://shuttle.nasa.gov/> (September 13, 1999).

National Aeronautics and Space Administration. *SpaceLink*. n.d. <http://spacelink.msfc.nasa.gov/> (September 13, 1999).

INDEX